PROJECT: STEAM

SCIENCE IN MOTION

LISA AMSTUTZ

Rourke Educational Media

rourkeeducationalmedia.com

Before & After Reading Activities

Before Reading:

Building Academic Vocabulary and Background Knowledge

Before reading a book, it is important to tap into what your child or students already know about the topic. This will help them develop their vocabulary, increase their reading comprehension, and make connections across the curriculum.

1. *Look at the cover of the book. What will this book be about?*
2. *What do you already know about the topic?*
3. *Let's study the Table of Contents. What will you learn about in the book's chapters?*
4. *What would you like to learn about this topic? Do you think you might learn about it from this book? Why or why not?*
5. *Use a reading journal to write about your knowledge of this topic. Record what you already know about the topic and what you hope to learn about the topic.*
6. *Read the book.*
7. *In your reading journal, record what you learned about the topic and your response to the book.*
8. *After reading the book complete the activities below.*

Content Area Vocabulary

Read the list. What do these words mean?

acceleration
conserved
drag
exhaust
friction
gravity
inertia
jet propulsion
kinetic energy
law
lift
mass
potential energy
vacuum
velocity

After Reading:

Comprehension and Extension Activity

After reading the book, work on the following questions with your child or students in order to check their level of reading comprehension and content mastery.

1. *What are Newton's three laws of motion?* (Summarize)
2. *Why is it important to wear a seatbelt when you ride in a vehicle?* (Infer)
3. *What two types of energy does a catapult demonstrate?* (Asking Questions)
4. *How can you see the laws of motion at work around you?* (Text to Self Connection)
5. *How is motion different in space?* (Asking Questions)

Extension Activity

Create an activity that demonstrates one of Newton's three laws of motion. Test it out on your classmates. Summarize what you learned.

TABLE OF CONTENTS

THE SCIENCE OF MOTION

In 1666, according to legend, Sir Isaac Newton was sitting under a tree when an apple fell on his head. *Ouch!* That got Newton thinking: Why do apples always fall toward the ground? Pondering this problem helped him formulate his **law** of **gravity**.

Twenty years later, Newton proposed three more laws known as the laws of motion. They help explain how and why objects move. We've all experienced these laws in everyday life, but Newton was the first to put them into words.

The law of gravity states that any two particles in the universe attract each other with a force that varies their **mass** and the distance between them.

REMAINING AT REST

The first part of this law seems logical; for instance, if you set this book down, it will stay put unless you give it a push or pick it up again.

Newton's first law is known as the law of **inertia**. Inertia is the tendency for objects to keep doing whatever they're doing. An object at rest remains at rest, and an object in motion remains in motion at the same speed and in the same direction unless another force acts on it.

Activity:

Try these activities to test Newton's first law for yourself!

An object at rest will remain at rest unless a force acts on it. That force must be greater than the forces keeping the object in place in order for it to move.

Gather:

1. plastic cup filled with water
2. sheet of construction paper
3. coin
4. index card

Do:

1. Set the cup of water on a sheet of construction paper.
2. Give the paper a quick yank. What happens to the cup?

3. Set the cup of water on a table.

4. Place an index card on top of the cup.

5. Place a coin on top of the card.

6. Flick the edge of the card. What happens to the coin?

ON THE MOVE

On Earth, **friction** and air resistance slow your skateboard down. If you could launch yourself into space, however, you could keep moving forever, because there is no air in space, and you would not encounter any friction to slow you down.

The second part of Newton's first law says that a moving object will keep moving unless a force stops it. If you hop on a skateboard and give it a push, you will move forward with the skateboard. If you hit a curb, the skateboard will come to a sudden stop—but your body will not.

Activity:

Build an egg-mobile to observe the effects of Newton's first law. Make sure to do this experiment outside and with adult permission. If your egg cracks, it will be messy.

This experiment shows what happens when a moving car hits an object. If passengers are not wearing a seatbelt, they can be thrown through the windshield.

Gather:

1. egg carton

2. wooden spoon

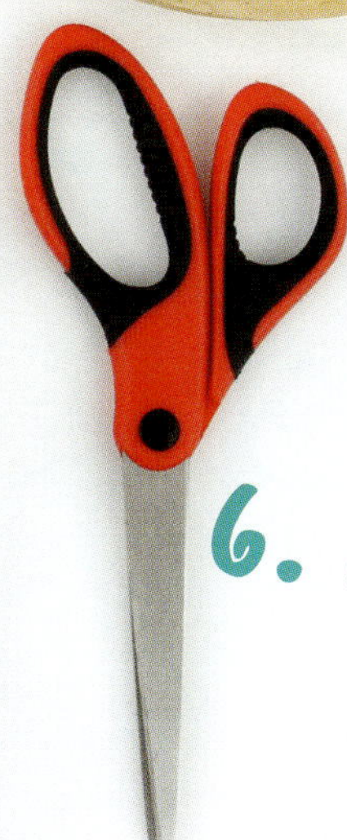

3. egg

6. scissors

4. small toy car

5. embroidery floss

7. board

Do:

1. Cut out one egg carton cup. Then cut off the top half of the cup.

2. Poke two holes near the bottom of the cup on opposite sides.

3. Tie the cup to the top of the toy car with embroidery floss.

4. Set an egg in the cup.

5. Lean a board against a chair or stool as a ramp.

6. Tape a wooden spoon to the floor about a foot (30.5 centimeters) away from the ramp.

7. Send the truck down the ramp. What happens when it hits the spoon?

8. Try to design a vehicle that will keep the egg in place when it hits the spoon.

MASS IN MOTION

If you kicked two soccer balls of different weights with the same force, which one would accelerate more quickly? If you said the lighter one, you are correct!

Newton's second law explains how force and **acceleration** affect the motion of an object. The greater the mass of an object, the greater the force needed to accelerate it at the same rate.

Activity:

How does mass affect motion? Try this simple activity to find out!

Force is equal to mass times acceleration. Scientists write this as F = m x a. The greater the mass of an object, the more force it takes to achieve the same acceleration. This is why a heavier car needs a more powerful engine to accelerate at the same rate as a smaller car.

Gather:

1. **thick book**

2. **tape**

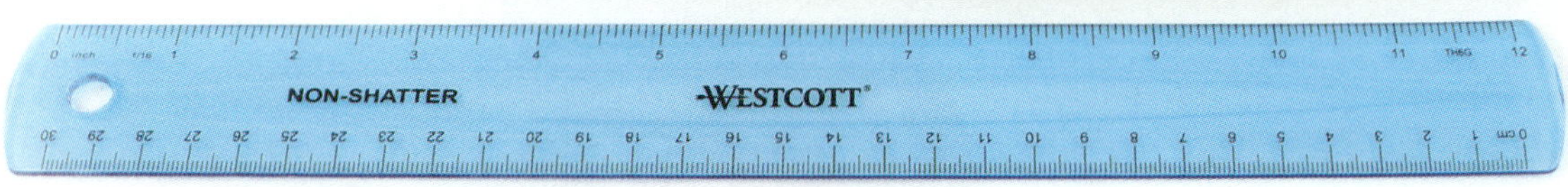

3. **flexible plastic ruler**

4. **several small balls (ping pong, golf, tennis, etc.)**

Do:

1. Place the book on a hard surface with its spine facing away from you. Hold the ruler against the spine. Let 6 inches (15 centimeters) of the ruler extend beyond the book.

2. Stick a piece of tape on the table about 2 inches (5 centimeters) behind the tip of the ruler.

3. Place a ball in front of the ruler.

4. Pull the ruler back to the piece of tape and release it quickly. Note how fast the ball moves.

5. Now try a different ball. Which was heavier? Which accelerated faster?

6. Experiment with balls of different weights.

TRY THIS

Place a skateboard on a hard floor and give it a push. Now set a heavy object on the skateboard and push it with the same amount of force. What happens?

ACTION AND REACTION

Newton's third law lets rockets move about in space. As material burns inside the rocket, hot **exhaust** leaves the rocket at a high rate of speed. This produces an equal reaction in the opposite direction, pushing the rocket forward. In this way, rockets can accelerate even in space, where there is nothing to push off against.

Sir Isaac Newton's third law is the law of action and reaction. It states that for every action, there is an equal and opposite reaction. When you push on a wall, for instance, the wall does not move away. Instead, it pushes back on your hand. If you hit the wall with a lot of force, the wall hits you back with the same amount of force. That reaction can be painful!

Squids and octopuses move by **jet propulsion**. They squirt liquid out of their bodies in one direction, which causes their bodies to move in the opposite direction.

Activity:

To watch Newton's third law in action, try out this balloon-powered boat.

Gather:

1. **Styrofoam egg carton**

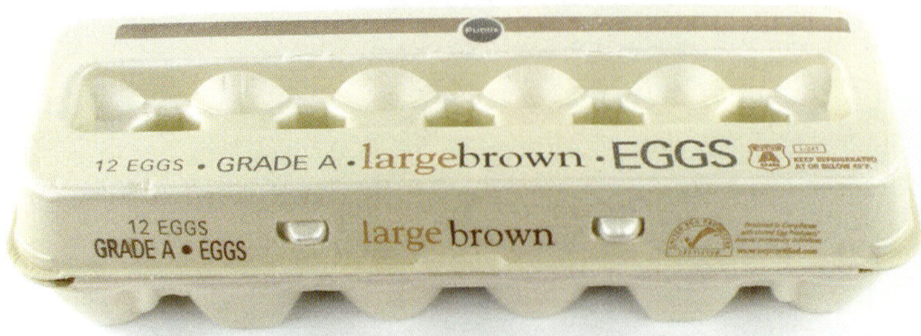

2. **plastic straw**

3. **small balloon**

4. **duct tape**

Do:

1. **Cut off the egg carton lid.**
2. **Poke a hole in the end of the carton.**

3. Insert the straw into the balloon and tape it in place.

4. Insert the straw through the hole in the lid and tilt it so the open end will be underwater. Tape the straw in place.

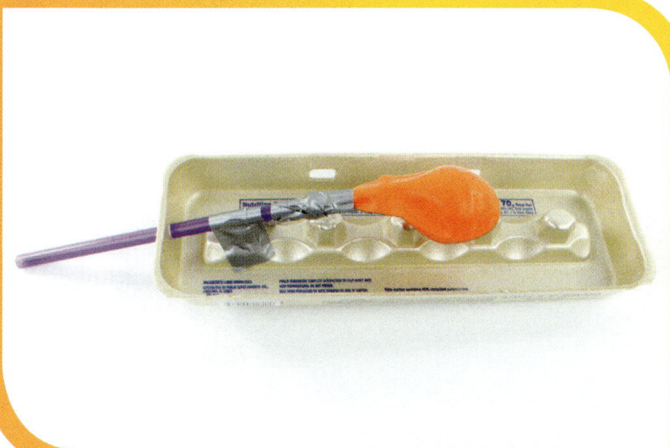

5. Blow through the straw to inflate the balloon. Then cover the end with your finger.

6. Place the boat in a tub of water. Release your finger. Watch your boat go!

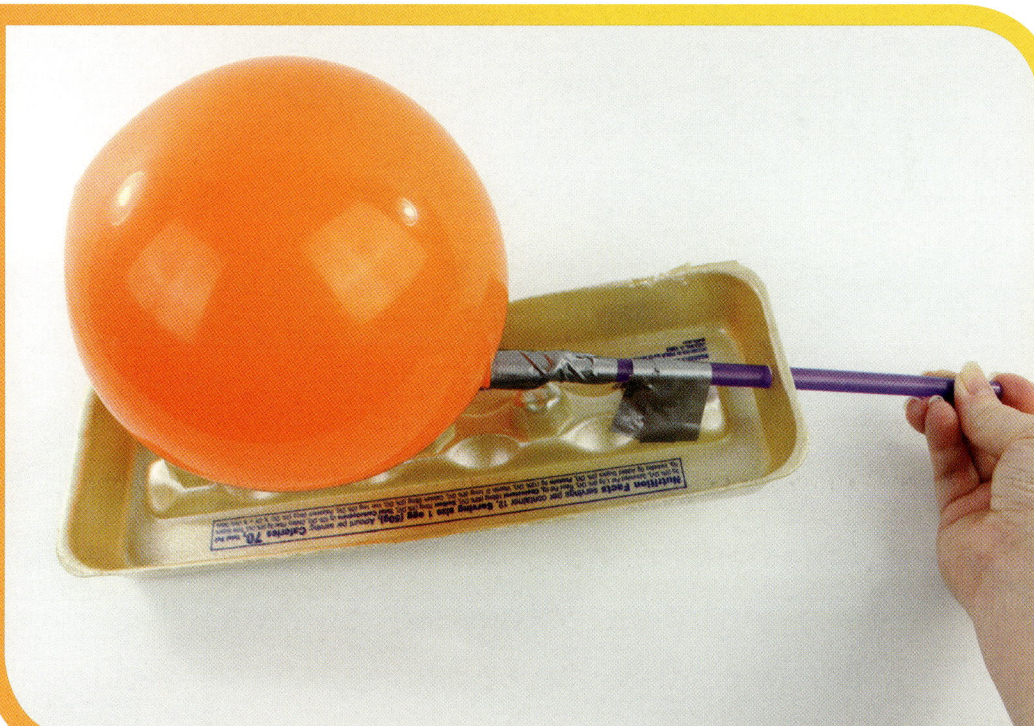

FALLING FAST

Newton's law of gravity tells us that things always fall toward the center of the earth. But did you know that all objects fall at the same speed, regardless of their size or mass? The exception is objects like feathers that can be significantly slowed by air resistance.

About 400 years ago, an Italian scientist named Galileo Galilei (1564 – 1642) concluded that objects should fall to the ground at the same rate in a **vacuum**. He tested out his hypothesis, so the story goes, by dropping objects from the top of the Leaning Tower of Pisa.

Activity:

Don't just take Galileo's word for it—you can test this idea for yourself!

Gather:

1. **metal cookie sheet**

2. **6 non-breakable objects of different sizes (balls, blocks, sticks, a feather, etc.)**

3. **chair**

TRY THIS

Hold a magazine in one hand and a piece of paper in the other. Drop both at the same time. What happens? Now put the paper on top of the magazine and drop it. This removes most of the air resistance. Did your results change? What would happen if you scrunched up the piece of paper?

Do:

1. Place the cookie sheet on the floor in front of the chair.
2. Stand on the chair and drop two objects onto the cookie sheet at the same time.

3. Which object hit first? Why? Try different combinations of objects to compare.

SLIP AND SLIDE

Have you ever tried to walk on ice? It's hard to move without falling! The smooth surface of the ice and the slight melting that happens when you press your foot on it make it almost frictionless.

Friction is a force that resists the movement of an object sliding or rolling against another object. Without it, we couldn't walk, drive, or do many other everyday activities.

Activity:

In general, the rougher an object is, the more force is needed to overcome the force of friction. Try using the force of gravity to move different objects down a ramp.

Gather:

1. **board**

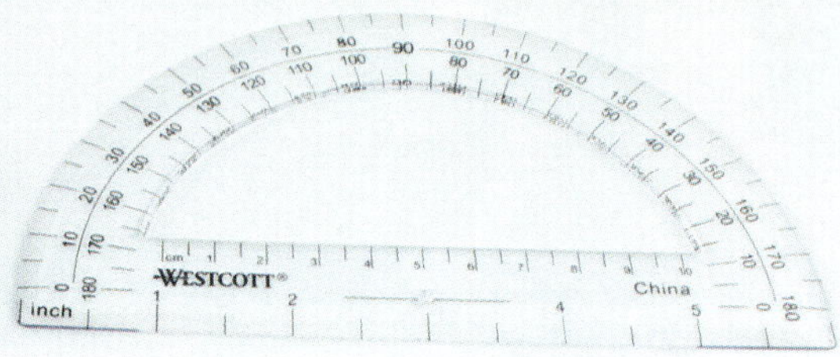

2. **several small objects (coin, eraser, wooden block, ice cube, key, etc.)**

3. **protractor**

4. **chair or stool**

Do:

1. **Lean the board against a stool or chair to form a ramp.**

2. **Use a protractor to measure the angle of the ramp as shown. Note the angle on a sheet of paper.**

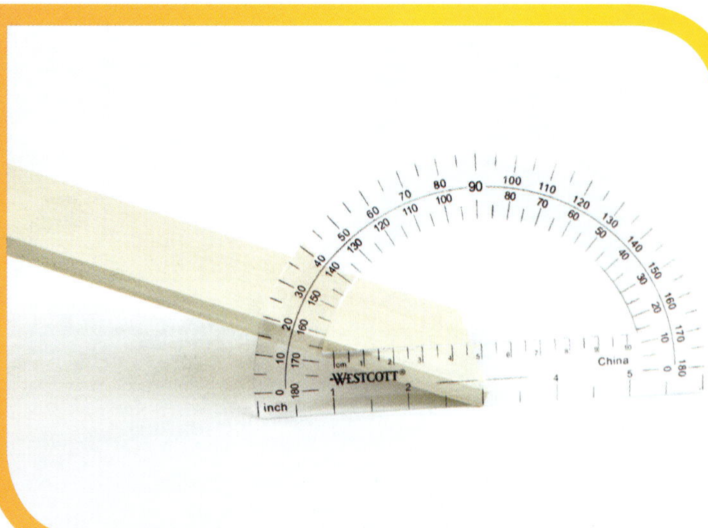

3. Set a coin at the top of the ramp. Did it slide down? If not, try a steeper angle. Find the smallest angle at which the coin will slide down the ramp.

4. Now try the other objects. Compare the angles needed to overcome the force of friction for each type of object.

5. What would happen if you used a more slippery board? What if you covered it with sandpaper?

WHAT A DRAG!

Have you ever stuck your hand out the window of a car when it is moving? It is easy to feel air resistance when you are moving fast. The air resistance is highest when your hand is perpendicular to the ground. Turning it parallel to the ground lowers the air resistance.

Although it is invisible, air is made up of a mixture of gases that take up space. When we push through these gases, they push back. The force of air molecules pressing against an object is called air resistance, or **drag**. In space, there is no air, so there is no drag. Scientists create similar conditions in the laboratory by removing all the air in a space to form a vacuum.

Activity:

Airplane designers work hard to minimize drag. An airplane's smooth curves let air flow around it easily.

Gather:

1. **notebook paper**

Airplanes need **lift** to fly. The shape of their wings—flat on the bottom and rounded on top—changes the speed and pressure of the air as it flows around them, creating lift.

Do:

1. **Try out several airplane designs.**

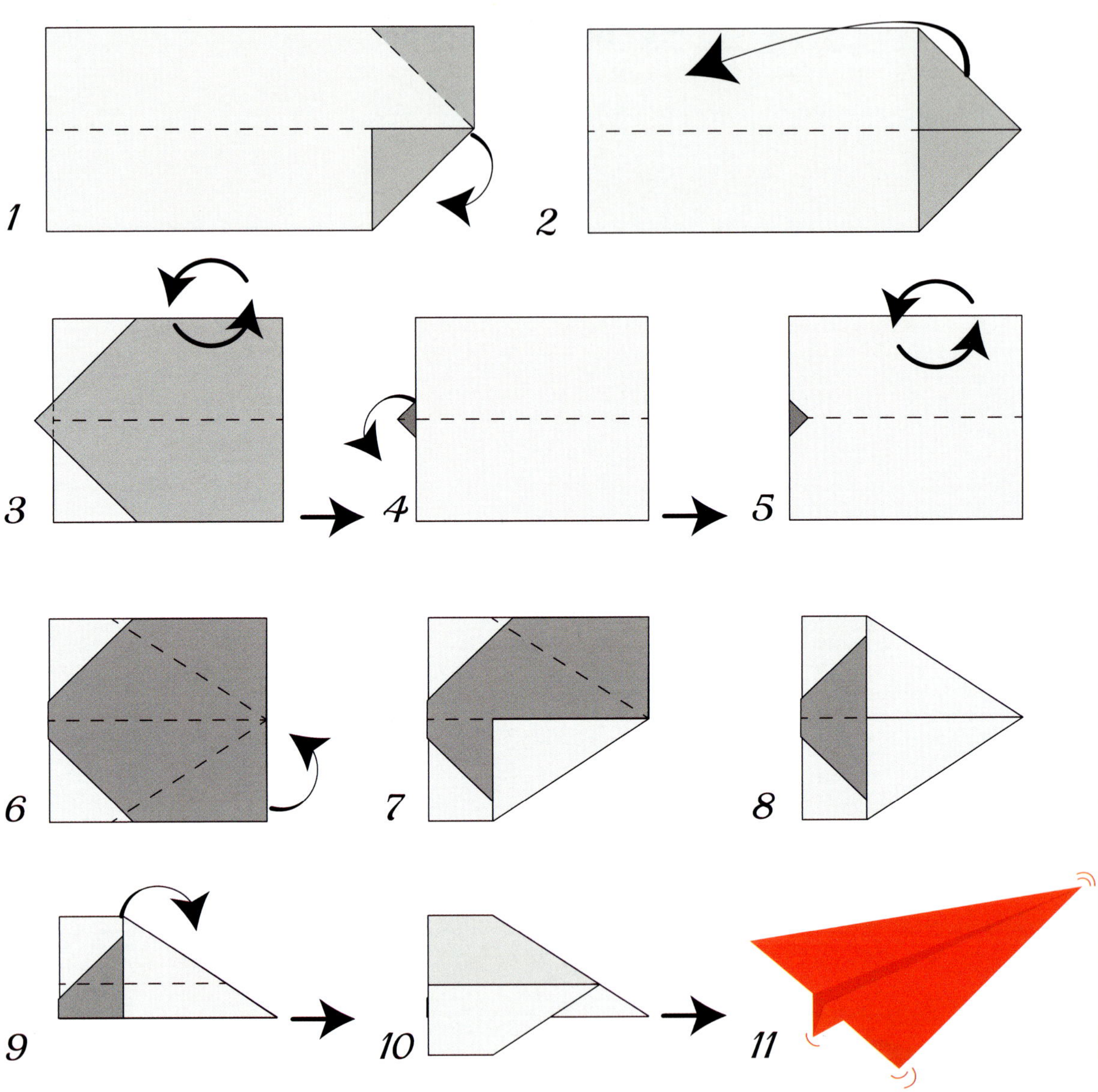

2. Standing in the same spot, throw each of the paper airplanes. Try to use the same amount of force.
3. Record the distance that each airplane flies.
4. Repeat your experiment several times and average the results.
5. Which plane flew the farthest, on average? Why do you think this is true?

ENERGY IN MOTION

For centuries, humans have been trying to create a perpetual motion machine. Once set in motion, such a machine would keep moving forever. It would have to be set in a vacuum (to eliminate drag), have no touching parts (to eliminate friction), and create no sound or heat (these are forms of energy too). So far, no one has succeeded!

Energy can be found in different forms in the world around us. When an object is moving, we say it has **kinetic energy**—the energy of motion.

Activity:

The amount of kinetic energy in an object is directly related to its mass and **velocity**. Experiment with balls of different mass to compare their kinetic energy.

Gather:

1. sand

2. box or pan

3. 4 different balls (e.g., marble, ping pong ball, golf ball, tennis ball)

Do:

1. Fill the box with several inches of sand.

2. Drop a ball into the sand.

3. Measure the crater formed where the ball hit the sand. Its size reflects the kinetic energy of the ball.

4. Repeat steps 2 and 3 for each ball. Make sure to drop each ball from the same height.

5. Create a chart to compare crater size and ball weight.

6. Try dropping the same ball from different heights. Does this affect the size of the crater? Why or why not?

FULL OF POTENTIAL

Wrap a rubber band around your thumb and pull it back with your finger. This stretching creates a type of potential energy known as elastic energy. When you release your finger, this energy is released—and the rubber band flies across the room.

Another type of energy is known as **potential energy,** or the energy of position. If you pick up a ball and set it on a table, the energy you used to pick it up is now stored in the ball as potential energy. It has the potential to fall, turning its potential energy into kinetic energy again.

Activity:

Observe potential energy in action with this catapult. The energy stored in the spoon when you press it down turns into kinetic energy when it is released.

Gather:

1. plastic spoon

2. hot glue

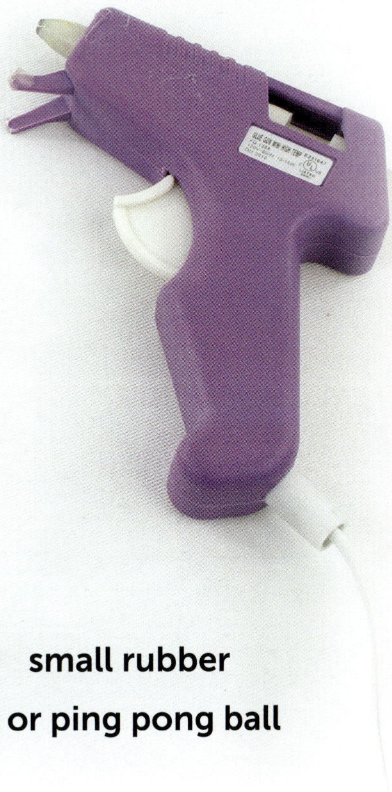

3. clothespin

4. 4 craft sticks

5. small rubber ball or ping pong ball

Energy is never lost; it just changes form. The law of conservation of energy states that in a closed system, the total energy of a system is **conserved**. If you set a ball at the top of a ramp, it has potential energy. When it starts moving, that potential energy changes into kinetic energy.

Do:

1. Clamp the spoon into the clothespin.

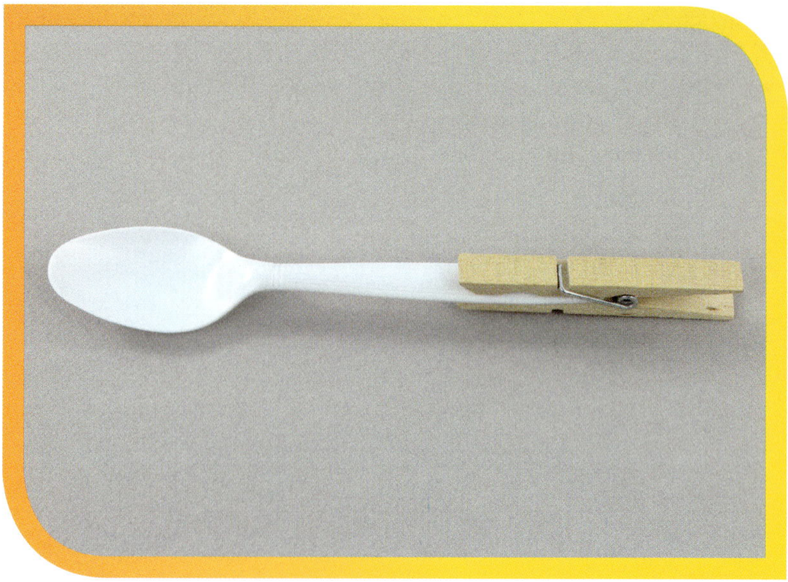

2. With an adult's help, glue two craft sticks into a V shape near the front of the clothespin.

3. Glue the other two craft sticks to the back of the clothespin as shown.

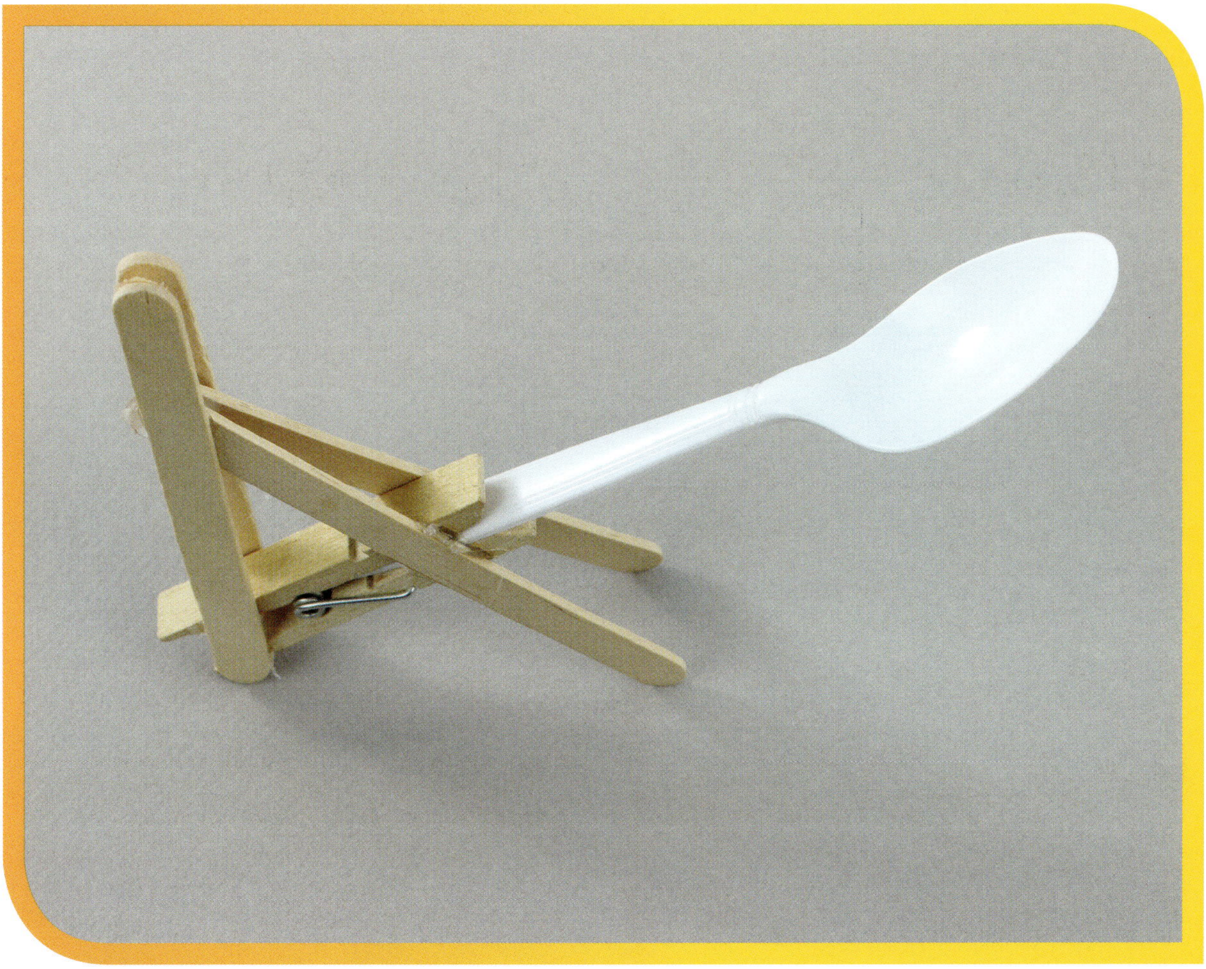

4. Set the catapult on a table and place the ball on the spoon.

5. Press down on the spoon. Then let go. Watch the ball fly!

Activity:

Store elastic potential energy in a spool racer by twisting the rubber band. Then let go and watch it turn into kinetic energy. Make several racers and race your friends!

Gather:

1. empty wooden spool

2. pencil

3. washer

4. paper clip

5. rubber band

6. tape

7. paint or markers

Do:

1. Decorate your spool with paint or markers.
2. Feed the rubber band through the center of the spool.

3. Feed the end of the paper clip through the rubber band.
4. Place a piece of tape over the paper clip.

5. Place the washer over the other end of the rubber band.

6. Put the pencil through the end of the rubber band and twist it tightly.

7. Set your racer on a hard surface. Release it and watch it go!

The science of motion affects us every day. Isaac Newton didn't invent gravity or the laws of motion—he was just the first to really understand and describe them. Today, scientists have a much better understanding of how these laws work, but there is still much to learn and explore. Maybe the next great discovery will be yours!

GLOSSARY

acceleration (ak-SEL-uh-RAY-shuhn): an increase in the rate of speed of something

conserved (kuhn-SURVD): unchanged

drag (drag): a force that slows the movement of an object through a liquid or gas

exhaust (eg-ZAWST): waste gases that are emitted from an engine

friction (FRIK-shuhn): the force that resists the rubbing of two objects against each other

gravity (GRAV-uh-tee): the force pulling things toward the center of Earth

inertia (in-UR-shuh): resistance to any change in motion

jet propulsion (jet pruh-PUHL-shuhn): the movement of an object by its reaction to a force pushing liquid or gas in the opposite direction

kinetic energy (ki-NET-ik EN-ur-jee): the energy of an object in motion

law (law): a statement about what always happens under certain conditions

lift (lift): an upward force on an airplane wing

mass (mass): the amount of physical matter that an object contains

potential energy (puh-TEN-shuhl EN-ur-jee): energy stored in an object

vacuum (VAK-yoom): a sealed space that contains no air or gas

velocity (vuh-LOSS-uh-tee): speed

INDEX

SHOW WHAT YOU KNOW

1. What does Newton's first law tell us?
2. Why do we need friction?
3. Why do some objects fall faster than others?
4. How do aircraft designers create lift?
5. What is the law of action and reaction?

FURTHER READING

Wood, Mathew Brendan, *Ballistics: From Rocks to Rockets with Science Activities for Kids (Build It Yourself)*, Nomad Press, 2018.
Challoner, Jack, *Maker Lab: 28 Super Cool Projects*, DK Children, 2016.
Sohn, Emily, *A Crash Course in Forces and Motion with Max Axiom, Super Scientist (Graphic Science)*, Capstone Press, 2016.

ABOUT THE AUTHOR

Lisa Amstutz is the author of more than 80 books for kids. She specializes in topics related to science, nature, and agriculture. Lisa's background includes a B.A. in Biology and an M.S. in Environmental Science. She lives with her family on a small farm in Ohio. You can learn more at her website: www.LisaAmstutz.com.

www.rourkeeducationalmedia.com

PHOTO CREDITS: Cover, 7, 8, 9, 11, 12, 13, 15, 16, 17, 20, 21, 24, 25, 28, 29, 35, 36, 37, 39, 40, 41, 42, 43, 44: © creativelytara; Page 4: © duncan1890; Page 5: © lucentius; Page 6: © man_kukuku; Page 10: © Warchi; Page 14: © Aksonov; Page 18: © 3DSculptor; Page 19: © TheSP4N1SH; Page 22: © Kierferpix; Page 23: © ventdusud; Page 26: © GorynVD; Page 27: © gbh007; Page 30: © Tijana87; Page 31: © subjug; Page 32: © graphicpoet; Page 33: © baona; Page 34: © Image Source; Page 38: © bradleym; Page 45: © kali9

Edited by: Keli Sipperley
Cover and Interior design by: Tara Raymo *www.creativelytara.com*

Library of Congress PCN Data

Science in Motion / Lisa Amstutz
(Project: STEAM)
ISBN 978-1-64156-464-9 (hard cover)(alk. paper)
ISBN 978-1-64156-590-5 (soft cover)
ISBN 978-1-64156-706-0 (e-Book)
Library of Congress Control Number: 2018930490

Rourke Educational Media
Printed in the United States of America,
North Mankato, Minnesota